TABLE OF CONTENTS

What Is Slime?

Oozy, gooey, sticky... it's slime! You might think slime is gross, but it is an important part of the animal kingdom. For their survival, animals use slime in wild ways. In this book, we'll take a look at how different animals make something sort-of-disgusting into something really amazing!

- Slime is not liquid, but it's not solid either. It's sort of in-between.
- Slime is made almost entirely of water. Each animal adds chemicals that it needs to use in its version of slime.
- Humans have several forms of slime, including saliva and the coatings of internal organs.

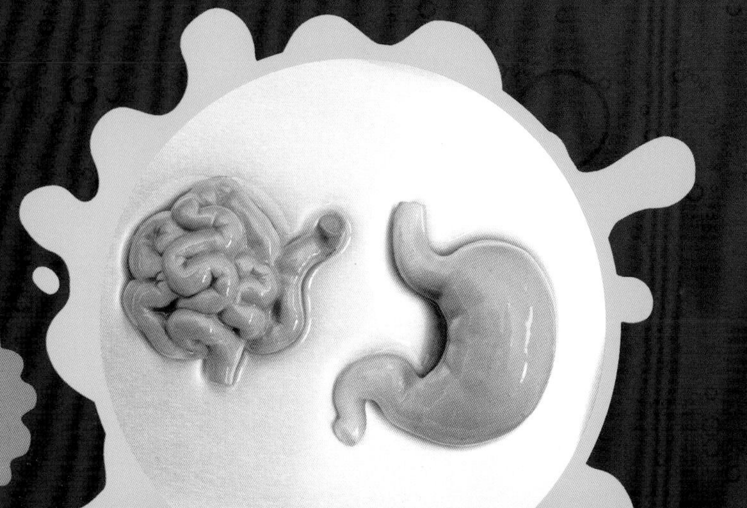

Slimy Bite
Science Word!

The scientific term for most forms of slime is a noun: MUCUS (MYEW-kus). If you see the word "mucous," that's an adjective and it means "slimy."

Slimy Bite

This giant tube of drifting slime that looks like a really, really long sock is super slimy. It's called a PYROSOME (PIE-roh-sohm). The pyrosome is also not just one creature, but millions and millions of little things called ZOOIDS. Each zooid is itself a tiny, slimy critter. Millions team up to form a single long, thin, plankton-sucking sea creature. The longest ever seen was 66 feet (20 m) long!

Slime Is Good!

Animals of all sorts use forms of slime to eat, breathe, hunt, and survive.

- In frogs and other **AMPHIBIANS**, a layer of slime helps keep their skin moist.
- Slime on fish skin protects fish from parasites (little critters that might eat the fish!).
- Slime on frogs and snails helps keep them from getting too hot.

When Slime Is Not Good

In 2017, drivers on a road in Oregon got slimed. A truck carrying hagfish to a fish market crashed. Hagfish (page 36) spew out buckets of slime when they are threatened. During the crash, they spewed like crazy! One car was almost covered in slime. The roadway had to be hosed off before anyone could drive. Gross!

That Is Just **Gross!**

You might eat slime but not know it. "Pink slime" is the nickname for a product made from what is left over after animals are butchered for meat. It is dyed pink and sold as part of ground beef or even pet food.

Pond Slime

Our first step into slime is all wet. Here's a look at the slime that covers freshwater ponds and the animals that live in it.

- Pond slime is mostly plant matter. The green stuff is **ALGAE** (AL-jee).
- Millions of tiny animals called **ZOOPLANKTON** can live in pond slime. They feed on the pond slime plants.
- Insects and fish feed on the zooplankton. They attract birds and other small animals ... and it all starts with slime!

Slimy Bite

More cool, tiny animals that live in pond slime are: protozoa, hydras, water fleas, copepods, and ostracods. Also, check out this picture of the super-tiny tardigrade. Isn't it cute?

That Is Just **Gross!**

Check your pond slime carefully. Some blue-green algae blooms can be deadly to pets and can make humans sick.

Frogs

That frog near the pond is not glistening because it's wet. It's shining because it's slimy!

- Frogs take in air (breathe) through their skin. A thin layer of slimy goo covers their skin to protect it. No breath, no life!
- Some frog slime has chemicals that prevent BACTERIA from growing on the frog's skin.
- All toads are frogs... but not all frogs are toads. Most toads are not slimy, but dry and rough on the outside.

Froggy Drink o' Water!

A very cool type of frog lives in Australia. The water-holding frog can take in enough water to survive a long, dry summer. It holds the water in a sort of bag or COCOON made of its own skin.

That Is Just **Gross!**

Frogs eat their own skin—slime and all! Ewww! When a frog sheds a layer of skin, it gobbles it up to make, well... more skin!

More Frogs!

Frogs are among the most famous slimy animals, so here are some bonus facts about our croaking friends.

Frogspawn

Some species of frogs use another type of slime. When the females lay eggs, they surround the tiny balls of future frogs with foamy slime. The "bag o' slime" protects the eggs as they float in the water. It looks like soap bubbles with eyes!

Foam Nest Frogs

Foam nest frogs use yet another kind of slime. They cover eggs with slimy foam away from the water. The slimy foam hardens, creating a safe place for the eggs to grow before hatching.

Slimy Bite

A thirsty frog doesn't open its mouth for a sip. It absorbs water through its skin. The slime layers help the frog get the water it needs by trapping it next to the skin.

That Is Just **Gross!**

How about a fly buffet? The holy cross frog from Australia spews out a super-sticky slime that covers its body when it comes out of **HIBERNATION**. Local flies looking for a quick snack get caught in the slime. Then the frog sheds its skin and slime to eat the flies!

Salamanders

Salamanders—the slimy critters that come in many colors! They look lizard-like, but they are related to frogs.

Salamanders are amphibians. They are born in water, but grow to live on land.

As with frogs, the salamander's slime is PROTECTIVE and helps it stay moist.

Look for salamanders near ponds, lakes, or streams. They need a body of water nearby to lay their eggs.

Slimy Bite

If a salamander loses its tail to a PREDATOR or an accident, it can grow a new one!

That Is Just **Gross!**

Don't touch that salamander! Especially the fire salamander. Its slimy skin contains poison. The bright coloring warns predators of the danger.

Snails

See that trail of slime on the ground? At one end or the other is a snail.

Snails are **GASTROPOD MOLLUSKS**, which are boneless animals.

Snails use slime to help them move. Their muscles squish them along. The slime makes traveling easier and smoother. This is especially useful when climbing trees or walls.

Snails can use slime like a door. They pull into their shells and build a slime barrier on the shell opening!

Slimy Bite

France is the home to a popular and slimy dish—snails. Chefs in some French and seafood restaurants cover them with garlic and butter and serve them up as escargot (ess-kar-GO).

Believe it or not, some people smear snail slime on their faces... on purpose. Some beauty experts say that a snail slime mask is good for human skin. However, not all scientists agree with them.

Slugs

Most people think slugs are just totally gross. We think that these slimy creatures are totally cool!

- Boneless and squishy, slugs are **INVERTEBRATES** called gastropods.
- Two slug **TENTACLES** are used for smell and vision. Another pair helps with the senses of touch and taste.
- Slug slime trails are like ID cards. Other slugs can smell the slime to track mates, or, we guess, look for friends!

Slimy Bite

Got slug slime on your skin? Don't try to wash it off. Slug slime absorbs water! One expert suggests waiting for the slime to dry. Then you can rub it off with your fingers.

That Is Just **Gross!**

Gastropods, including slugs, trail only insects as the most numerous animals in the world. That's a lot of slime!

More Slugs!

Super Slime!

Predators tracking the red triangle slug in Australia are in for a surprise. This slug oozes out slime that becomes super sticky. The predator gets stuck and sometimes can't move for a day or more. One scientist found a frog stuck that way. The little froggy was slimed so well, the scientist carried the frog—stuck on a stick—back to his lab!

Down It Goes!

Slugs use slime to move easily on the ground. They can also spew out a strand and use it like a rope. They might lower themselves to another branch or from a tree to the ground. Tarzan slime?

That Is Just **Gross!**

Slugs eat their own slime. They also eat other slugs' slime. Yuck!

Banana Slugs

Bright yellow and longer than your finger, the banana slug loves to come out in the rain!

- Banana slugs are gastropod mollusks; they have no bones! Growing up to 10 inches (25 cm) long, they are the second-longest slugs in the world.
- They ooze out sticky slime that helps them climb redwood trees, and makes them taste nasty to predators!
- They live in very wet forest areas on the northern west coast of the United States and in Canada.

Slimy Bite

The sports teams of the University of California, Santa Cruz, are ... the Banana Slugs! Sammy the Slug appears on signs on campus. Slugs are much cooler than the typical team mascots like tigers, bears, or eagles!

That Is Just **Gross!**

A tradition on a trip to Northern California forests is to find a banana slug... and KISS it! Ewww!

Protective Slime

Slime is gross, but it can also save an animal's life. Frogs and slugs are covered with slime that helps them breathe and get water. Other animals use "safety slime" in wilder ways.

Parrotfish

Say you're a nice little parrotfish and it's time for a nap. How do you keep the tiny critters called **GNATHIIDS** from nibbling on your tasty skin while you take five? You cover yourself in slime. Parrotfish surround their resting selves with a layer of goo that keeps out the nibblers.

Possum

Possums are mammals often found in forests ... or backyards! They are **PREY** for larger animals like foxes, wolves, or bears. When attacked, possums flop over and pretend they are already dead. They also ooze out a really disgusting, slimy goo. Dead and stinky ... not a great meal!

Swiftlet

How'd you like to live in a house made of spit? Cave swiftlet chicks have no choice. That's how their bird parents make a nice place for eggs. The birds use spit to form the nest. The swiftlet's spit is like concrete; it starts out gooey but hardens quickly. The hard shell protects the eggs as they wait to hatch.

From where do possums ooze their defensive slime? Well...from their butt, actually.

Lungfish

Meet these ancient, slime-spewing, mud-dwelling animals... that take really, really long naps!

- Lungfish all live in fresh water, not the ocean, and can survive out of the water. A special type of lung helps them breathe air.
- They live in muddy pools. When the pools dry in hot weather, the lungfish hibernate.
- Here's where the slime comes in: They spew out thick goo to form a sort of cocoon under the mud.

Slimy Bite

Lungfish are ancient! Fossils more than 400 million years old look like today's lungfish! One of their modern cousins is the coelacanth, one of the world's oldest fish.

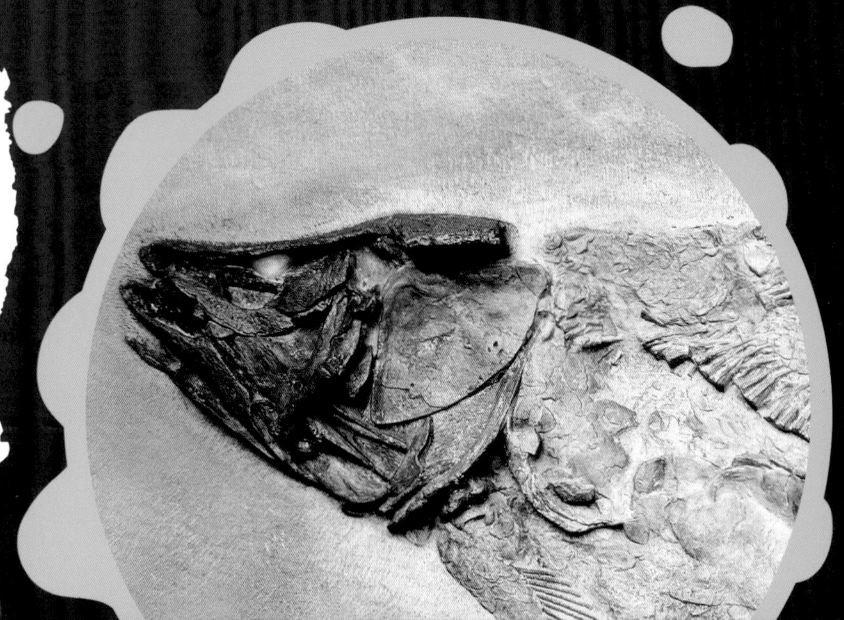

That Is Just **Gross!**

Lungfish can basically shut down most of their body systems and live in their slimy cocoon for up to three years!

Jellyfish

No bones, no brain, no heart ... no problem! Gooey, sticky, blobby jellyfish have been doing fine for 500 million years.

- Talk about gooey: Jellyfish bodies are made from 95 percent water!
- Some jellyfish are BIOLUMINESCENT. That's a long word that describes when an animal has body chemicals that glow.
- Don't pick 'em up! If you find a gooey dead jelly on a beach, don't touch. Many species have stinging tentacles.

Slimy Bite

Jellyfish are blooming around the world ... and that's not good. In Israel and Scotland, nuclear plants were shut down when jellyfish clogged their pipes. In Japan, enormous jellies, called Nomura's jellyfish (they can be 485 pounds/220 kg), grew so numerous that fishing boats could not do their work. When jellyfish gather in huge numbers, it is called a "bloom."

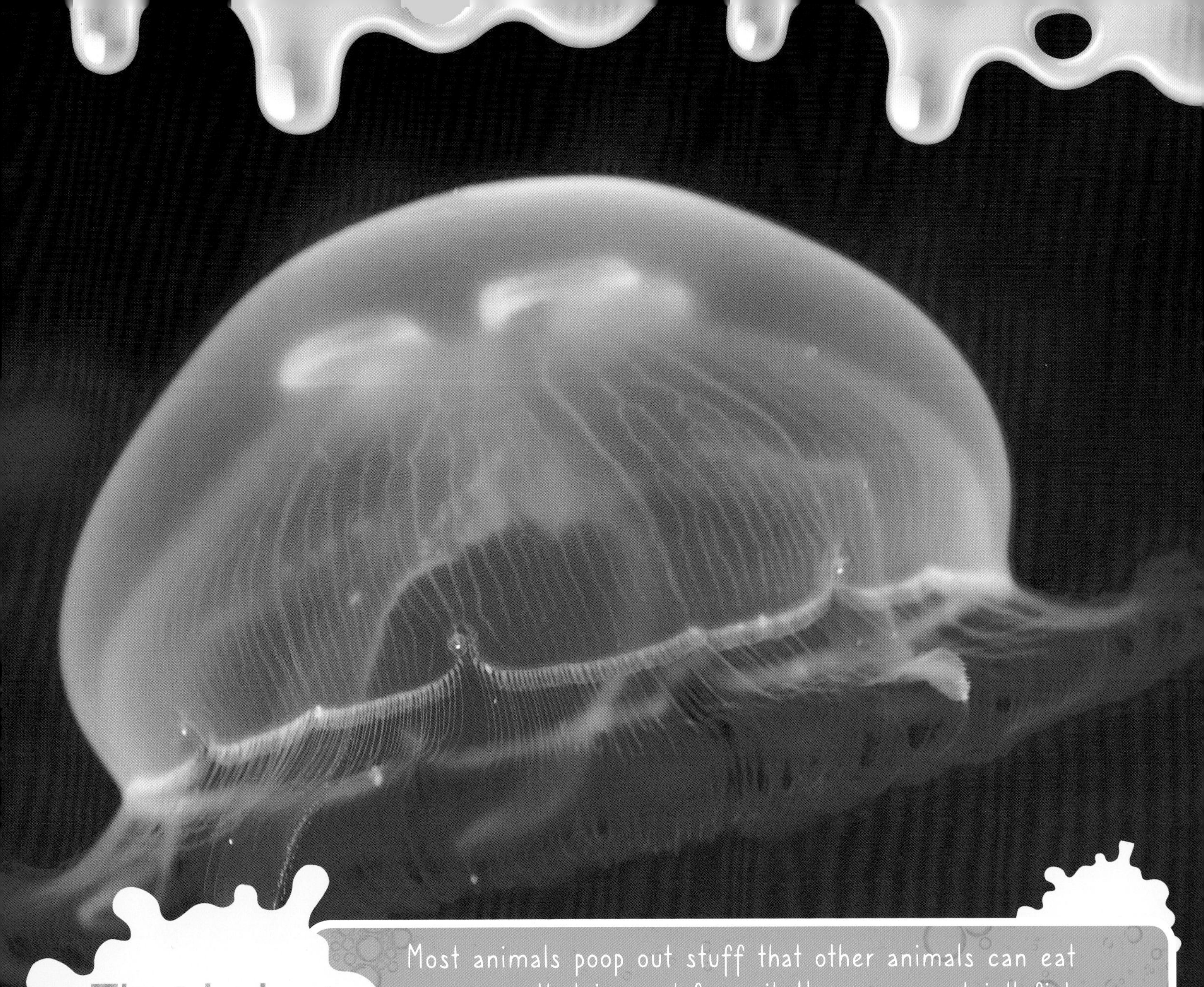

That Is Just **Gross!**

Most animals poop out stuff that other animals can eat or use, or that is good for soil. However, most jellyfish EXCRETE (fancy word for pooping) stuff that other ocean animals can't eat. Some scientists worry that if there are too many jellyfish, the ocean will be filled with inedible food. They call such a future "the rise of slime!"

Sea Cucumbers

Don't look for these "cukes" on vines like veggies. Dive to the bottom of the ocean and prepare for a slimy experience.

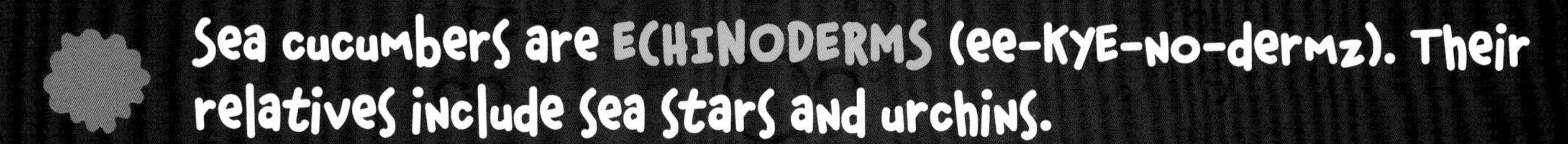

- Sea cucumbers are **ECHINODERMS** (ee-KYE-no-dermz). Their relatives include sea stars and urchins.

- These are sort of mysterious creatures. Scientists don't know how long they can live or how big they can get. Sea cucumbers are also the only animals in the world that can eat through their lungs!

- More than sixty species are **ENDANGERED**.

Slimy Bite

New thing not to want to be: Pearlfish. These tiny creatures live inside, well... a sea cucumber's butt. Sea cucumbers help keep ocean floors clean. They take in sand, suck out the food, and poop out "clean" sand. Thanks, guys!

That Is Just **Gross!**

Sea cucumbers can spit out all of their internal organs, which are sticky and used for defense ... but don't worry, they grow back.

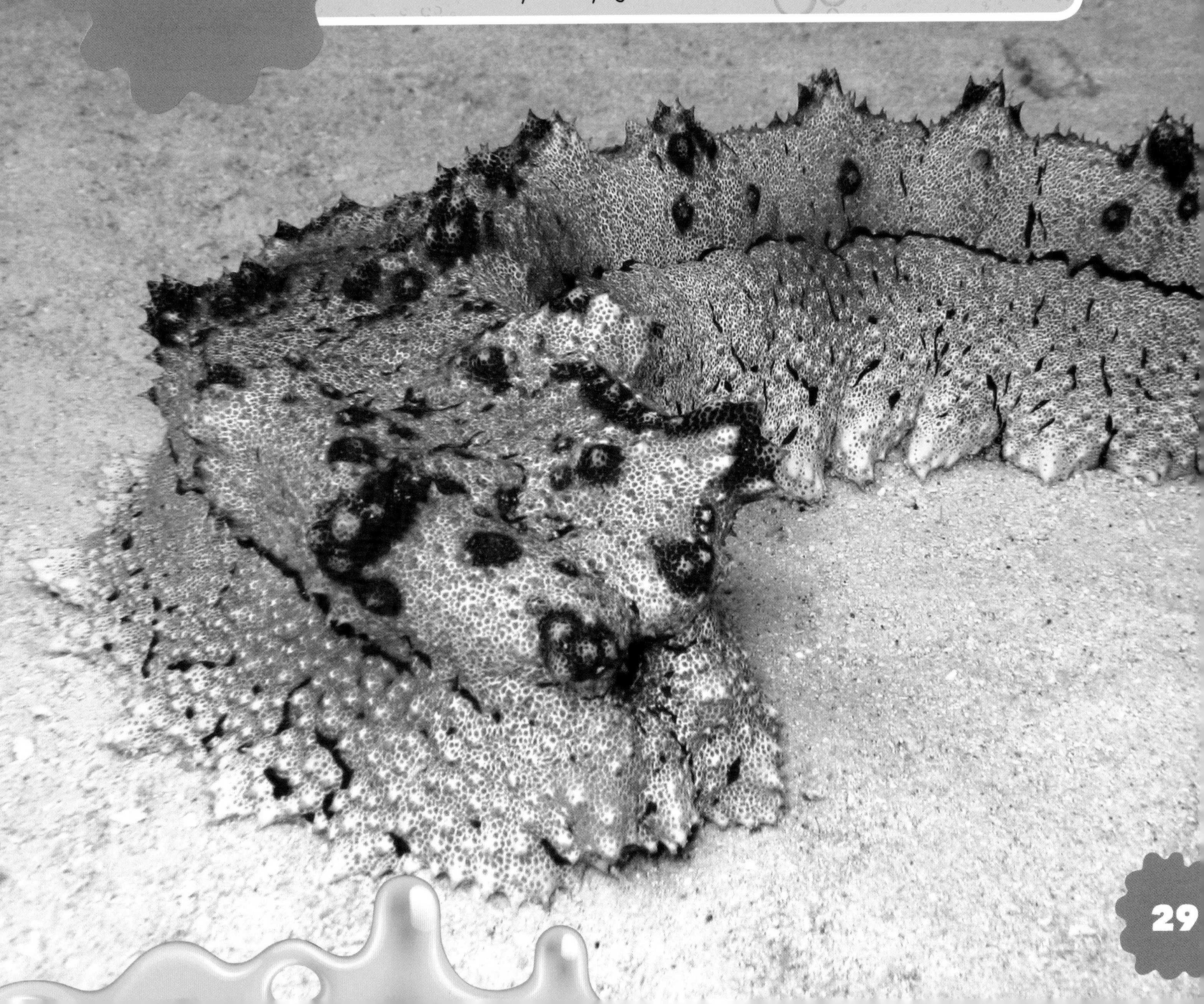

Sea Hares

Outside? Slimy and gooey. Inside, packed with "don't touch me" colored ink!

- Sea hares are a type of sea slug, an underwater mollusk.
- They can crawl along the sea floor or swim. But, no, they cannot hop!
- Sea hares don't need a mate. Each animal is both male and female. A sea hare can lay 80 million eggs at one time, then die after laying the eggs.

Slimy Bite

Pick up a sea slug out of the water and you'll feel its slimy outer coating. Try to grab one underwater and you might get inked! Sea hares spit out foul-tasting, poisonous colored ink. The color depends on what type of algae they eat.

That Is Just **Gross!**

Most sea hares are not much bigger than your hand. Some species, though, can be as long as 16 inches (41 cm)! That would be one long, gooey glob to try to hold! The big ones weigh as much as a medium-sized dog.

Limpets

Goo like glue, slime like bread crumbs, and a name like a bomb—limpets are pretty cool!

- Limpet teeth are stronger than spider silk, once thought to be nature's strongest material. The teeth actually contain a super-hard mineral called goethite.
- A limpet that is born a male can change into a female later in its life.
- The name "limpet" is also given to a type of weapon—a small bomb that attaches to an enemy ship with magnets.

Slimy Bite

Limpet slime is like bread crumbs. Remember Hansel and Gretel, who left bread crumbs to help find their way out of the forest? Limpets leave a trail of slime when they move across rocks so they can find their way back to their favorite spot!

That Is Just **Gross!**

Limpets use a type of slime like glue. They ooze it out and use it to stick very firmly to rocks near the shore.

Blobfish

C'mon ... smile, fish! Slimy blobfish, living near Australia, probably aren't as grumpy as they look.

- Lazy? Yup. Most blobfish just wait for food to swim by.
- These poor guys were once named the world's ugliest animal. That's just mean!
- Blobfish don't lie on the ocean floor. Their boneless bodies float just above it.

Slimy Bite

The blobfish became world-famous when the one shown on page 35 was dragged up from the sea near New Zealand. His smiling face went viral. People called him "Mr. Blobby"!

That Is Just **Gross!**

Author Michael Hearst called the blobfish the "Jell-O® of the sea" in a goofy song.

Hagfish

These hard-to-handle animals are like underwater slime fountains!

- Holes in the hagfish's sides spew out slime if it is attacked. Lots of slime. The slime can also be used to paralyze prey.
- These fish (nicknamed slime eels) are mostly boneless, but they do have skulls and teeth.
- Celebrate Hagfish Day if you want to. It's the third Wednesday of October.

Slimy Bite

Online videos show people on a fishing boat picking up huge globs of slime from hagfish in a bucket. It looks like they are picking up giant handfuls of snot.

That Is Just **Gross!**

What is the hagfish's favorite meal? Dead, rotting fish!

Slime That Helps

Scientists are studying the world of animal slime. Yes, that is kind of a gross way to spend your day, but it might be worth it. Slime might be able to help people in some very interesting ways.

Anti-Plastic Jellyfish

Jellyfish are basically giant balls of gooey slime. In one experiment, scientists found that jellyfish slime could suck **NANOPARTICLES** of gold out of water. (A nanoparticle is a piece of something so small that it is microscopic.) Ocean water is, sadly, filling up with nanoparticles of plastic. One group of scientists in Slovenia is experimenting with jellyfish slime to see if it can be used to filter out the plastic! The slimy work goes on!

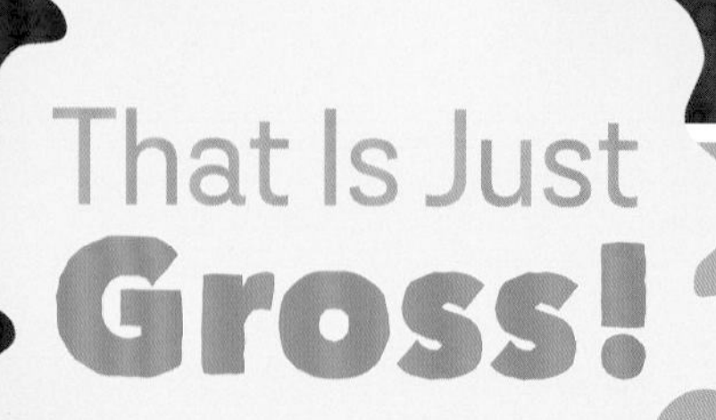

"There is lots of nasty work and the smell is not pleasant. But everything for science!" says jellyfish scientist Katja Klun. She has spent hours collecting jellyfish slime in buckets!

Got the Flu? Get a Frog!

The flu is caused by a virus. Scientists are looking at a particular type of frog whose slime has stuff in it that... kills flu virus! We're a long way from forcing frogs (specifically, the fungoid frog of India) to give up some slime to make medicines that can cure the flu. But how cool that it might happen some day? Thanks, frogs!

I ♥ Slime!

You know how sticky bandages don't stick too well to wet skin? A type of slug, called the dusky arion, might help with that. Harvard scientists have made a type of surgical glue with help from that slug's slime. The glue can make bandages that can be used (they hope) on wet places like hearts.

Ribbon Worms

You can't tie a bow with these ribbons, but you can get pretty grossed out by them!

- Slime 1: Ribbon worms spread one type of slime along their bodies. The slippery goo helps them slither easily along ocean floors and through underwater mud.
- Slime 2: Some species have poison slime on their tongues. Cover a crab with it, wait a moment, and ... dinner!
- The bootlace species of ribbon worm can be more than 98 feet (30 m) long!

Slimy Bite

Like its wormy cousins on land, the ribbon worm can grow back parts of its body that get snipped off by a predator.

That Is Just **Gross!**

When attacked, ribbon worms can spit out their long tongues and leave them behind, hoping their attackers think the tongues are worms!

Glowworms

Not really worms, these baby insects use slime to go "fishing." They live only in a few caves in New Zealand.

- Fungus gnats are insects. Their babies are called LARVAE. The gnats lay eggs that hatch into larvae and live on a cave roof.
- Before they become adult gnats, the larvae dangle lines of slime that are then covered with drops of more slime.
- Then they light up the lines and drops with bioluminescence. The larvae glow! Tiny mayflies get trapped in the slime drops and become larvae food.

Slimy Bite

The glowworms in the Waitomo Caves were first seen by people in 1887. The local Māori Chief, Tane Tinorau, led an English cave explorer to the spooky light. Today, Chief Tinorau's DESCENDANTS guide visitors from around the world on small boats into the glow.

That Is Just **Gross!**

The slim slime tubes that hold the glowing droplets are like long pieces of spit, straight from the larvae's mouths.

Velvet Worms

Ready... aim... slime! This little rainforest worm has a superhero talent!

- Velvet worms eject streams of slime onto prey. The slime hardens and traps the animal.
- They move on tiny legs located on almost the whole length of their body, sort of like caterpillars.
- Velvet worms have a second slime. Their skin is coated with it to keep in moisture.

Slimy Bite

More than 180 species of velvet worms have been found. They come in amaz... colors, including pink, purpl... and red, along wit...

That Is Just **Gross!**

The **GLANDS** that make slime for the velvet worm make up more than 10 percent of its body weight. How much would your slime gland weigh at that rate? Do the math!

Poison Slime

Slime is just gross, of course. But some slime is more than gross... it's deadly! A few animals produce slime that they use to defend themselves against predators. A hungry animal looking for a snack might just get a mouthful of poison! Read on about some animals you definitely want to avoid!

Poison Dart Frogs

There is a reason these South American frogs are so colorful. They are telling predators, "Hey, dude! If you eat me, you'll die!" Good advice. Native people have tipped arrows with poison from these animals' backs. They use the arrows to hunt larger animals.

Slimy Bite

Why does the poison slime not hurt the animals that have it? Science word: ADAPTATION! Animals that use slime for defense have adapted (or changed over time) so that they can have poison on or in them but not be harmed by it.

These squishy animals might look like a nice chew toy for your dog, but keep Rover far away! When attacked, cane toads ooze out a white goo that is very poisonous. The goo can kill small animals and make humans very sick. In January 2019, thousands of cane toads invaded an area near Palm Beach, Florida. They are also a big problem in parts of Australia.

Komodo Dragons

If you don't brush your teeth, you could end up like this enormous lizard. This animal's nasty slime comes from inside its mouth. And the slime is poisonous because the komodo's mouth is packed with disgusting bacteria it gets from eating rotten meat!

GLOSSARY

adaptation: a change made over time to aid an animal's survival

algae: simple plants that have no leaves or stems and grow in or near water

amphibians: animals that can live on land and in the water

bacteria: microscopic living creatures, often made of just one cell

bioluminescent: able to glow using body chemicals

cocoon: a container made by an animal for protection or metamorphosis

descendants: animals or people that come from earlier related animals or people

echinoderm: a type of marine animal; most have an outer shell and very simple digestive systems

endangered: living with the chance of disappearing completely

excrete: to push out of a body

gastropods: small invertebrate mollusks such as snails or slugs

glands: body parts that let out chemicals that help with body systems

gnathiid: a tiny marine crustacean

hibernate: move to a dormant state by slowing down body systems and staying protected for an extended period of time

invertebrate: an animal without a spine

larvae: the young of insects

mollusk: a type of invertebrate animal that has an outer shell

mucus: a slippery, gooey substance put out by living things

nanoparticle: a tiny object that is measured in nanometers; there are 2.5 million nanometers in an inch!

predator: an animal that hunts other animals

prey: an animal that is hunted and eaten by other animals

protective: covering from harm

pyrosome: a tube-like marine structure made from many, many tiny individual animals

tentacles: arm-like extensions of an animal's body

zooid: a small, multicellular animal that combines to make larger living structures

zooplankton: microscopic marine animals